TO WIN A WAR

(The Jewish Way)

~ A guide for post-war sanity ~

Rabbi Manis Friedman
and Dr. Elad Ben Elul

For information, contact:

It's Good to Know Publishing
Brooklyn, New York
ItsGoodToKnowPublishing.com
Email: info@itsgoodtoknowpublishing.com

Layout and cover design by Zalman Friedman
Printed in the United States of America

ISBN 979-8-9899444-0-8 (trade paperback)

Note to the reader: Please note that, in the interest of clarity, some of the quotations included in this book are either paraphrased or loosely translated but are nonetheless faithful to the original intent.

To all the
heroes
of the
Israel
Defence
Forces

CONTENTS

FOREWORD

Woke and Unwoke

October 7th, 6:30am. Shabbat morning. Tel Aviv. I awaken to the sound of an alarm, and run to the shelter. After a few moments I wonder—should I stay home, or leave to go to the synagogue? I decide to go. It's Simchat Torah, after all. As a child I never celebrated this holiday, but today it feels so essential, I couldn't bear to miss it.

As we sang and danced with the Torah scrolls, the news began to trickle in. Whispers, worried faces, men in IDF uniforms hurrying out the door. And a strange, almost embarrassing blend of joy and fear, inspiration and shock.

By evening the entire community was hiding in shelters, having heard the horrific news: a genocidal attack on our brothers and sisters, in our own homeland. From that point forward nothing would remain the same.

As an anthropologist I was trained to promote the rights of indigenous peoples and ethnic minorities around the world. The Western Empires, I was taught, colonized people of colour and took over their ancestral homelands through violence, exploitation, and cultural appropriation. In this Academic milieu it was made very clear who are the bad guys and who are the good guys. Regarding Israel, there was no question that the Palestinians are the persecuted indigenous minority, and the Jews are the Western colonizers. Under these strict principles, in this political climate, I could never have imagined taking part in the creation of the book you are about to read.

Only when I realized what a Jew is and who the real natives in the story are was I able to think independently and see reality for what it is. Only when God entered the picture was I able to make sense of my role as an Israeli Jew, and my duty in continuing the mission of my ancestors. It was not a question of Left or Right, but a matter of truth, courage, and faith. Political correctness and academic fantasies have now become not only foolish, but dangerous. It was time to speak the truth.

When I started listening to Rabbi Manis Friedman, one of the many wonderful things that happened in my life was that I discovered a new language, in which I can speak about morality, Judaism, and Godliness with full confidence. The Rabbi's clear, unapologetic, and insightful approach makes truth accessible again. The fear of being politically incorrect fades away when you know that you are in the service of God.

Ever since the war began, Rabbi Friedman and I have been giving open Zoom lessons to Israelis, guiding them on the most pressing issues of our time. Every week at least 200 people log in to our Zoom session, waiting to hear a word of encouragement, a bit of wisdom to carry us through these dark, confusing days. As many have told me, these sessions have kept us optimistic. Relaxed. Sane.

It was then that we decided to write this urgent book, by conducting long video interviews and writing into the night. Our intent was to organise the most relevant and essential information that a Jew might need in these challenging times. This book addresses some of the key principles that pertain to all wars. Yet it could never have been published before October 7th. The world simply wouldn't have comprehended it (and I would not have dared write it). But now it appears to me to be the clearest, most straightforward portrayal of a quintessentially Jewish approach to war that anyone could want. I hope you will share it with your friends, family, and colleagues.

If you know people who have influential roles in government, in the media, and in the military, please do not hesitate to send them this book. I am certain that it can help make some very complicated issues astoundingly simple.

Best wishes,
Dr. Elad Ben Elul

INTRODUCTION

Who Doesn't Want to Be Moral?

This book is being written in a time of war. Its purpose is to support, educate, and deepen the understanding of all who are affected by the war, as well as those who are simply interested in learning more about its meaning and implications.

Throughout the book we will discuss the history of the region, clarify the relevant terminology, assess the political and ethical framework, and perhaps most significantly, offer some insight into what the very near future portends. The various topics we'll discuss might be analysed from many different angles and disciplines, but we will focus on the spiritual/Chassidic perspective.

Some might be tempted to read this book for the

purpose of arguing in favour of Israel, winning social media battles with anti-Israel protestors, or convincing friends and colleagues as to the legitimacy of our nation and its actions. We would advise against this. It should not be the focus of your energy and efforts. By now, anyone who argues that Israel is an "occupier" in its own land, or that the indigenous residents of the land of Israel are in fact the Palestinians, is beyond the reach of reason. In other words, this book is *not* a polemic. It's not our aim to critique or attack other people's opinions about Israel, or about the Jews. Rather, our intention is to present a coherent, Torah-based approach to statehood, and, unfortunately, to war.

Nowadays everybody wants to be moral, and just about everyone is convinced that he is in fact being moral. But most have no idea what that means. Instead of educating themselves, they take sides the easy way: "Who was here first? Who is weaker? Who is poorer? Who has more casualties?" With this childish approach to morality, the academic and journalistic "debate" is flawed at its core. And so, we believe that the healthy approach is to address those who are genuinely interested in pursuing justice and truth — and bring clarity to the conversation based on principles rather than prejudice.

As an emissary of the Lubavitcher Rebbe in Minnesota I was working with students during the Yom Kippur war in 1973. At that time Arabs were holding Israeli hostages, and the students asked me to help them organize a petition to gain their release. We stood in the street freezing, collecting signatures, until I couldn't take the cold anymore and went to sit in my car. When I turned on the radio (and the heater) I heard that the hostages had already been released the day before. And I became curious as to how that might affect the students' feelings about their efforts. The point of the story is that our personal need to feel moral and righteous is not necessarily consistent with the facts of life. If morality is here to serve *me,* it's no longer focused on the truth.

Hopefully, this book will help us re-discover morality in its Godly context. May we hear good news from Israel and see great miracles ahead!

Rabbi Manis Friedman

Your Birthright: Israel as the Promised Land

Israel's Divine Purpose

Ever since 1948 the world has been debating whether Jews have a right to live in and establish a sovereign national presence in Israel. As Jews, we often try to stake our claim and justify ourselves based primarily on historical precedent: *we were persecuted and murdered in other places; the British Empire allocated the land to us; we agreed to share the land as partitioned, but the Arabs did not;* and so on. These strategies, intelligent as they may be, generally come off as defensive, and haven't proven very useful. Let's try looking at it from a different standpoint.

How have the nations of the world come to own their lands? How did they acquire such vast territories? They had to conquer them with their swords. The lands were never "theirs" to begin with. Every single country in the world, without exception, began by conquering its territory. In that sense, the land of Israel belongs to the Jews at least as much as any other country belongs to its conqueror. But there's more.

History shows that there is no land on earth that is as deeply and foundationally associated with one people for such a long time as Israel. No matter which empire or occupying army comes along and subjugates its indigenous people, it remains "Eretz Yisrael." The Babylonians, the Greeks, the Romans, the Turks, the Brits – they all tried in vain to claim this land by sending us into exile and/or changing the name of the region ("Palestine," for example.) Yet every child knows, in a way that has nothing to do with "International Law," that Israel is the land of the Jews. Why is it that every oppressor is able to swallow up the lands they defeat in battle, but none can ever take true and lasting ownership of Israel?

It's generally accepted that anyone can leave the religion of their birth and upbringing and convert to a different one. But Torah tells us that Jews are the exception to the rule. That's because the Jewishness of the Jew is an essential fact of life, a God-given condition. In a similar way, though the world at large believes that when it comes to land, whoever conquers it owns it, the holiness of the land of Israel is inextricably bound together with the holiness of the Jewish people. The Jewishness of Israel, in other words, is non-negotiable, and stands above choice or reason. This truth is so deep and certain, it surpasses and precedes creation itself.

In the Torah we learn that God repeatedly promised the land of Israel to our father Abraham and his Jewish descendants—to Isaac, to Jacob, to Moshe, to Joshua, and so on and so forth. But our right to Israel begins even earlier than that. After the great flood subsided, God divided the once-again inhabitable earth among Noah's sons. Shem received Israel, Ham received Egypt (Africa), and Yaphet received Europe. This was the very beginning of civilization. Shem did not have to conquer Israel, because it was granted to him long before the land was repopulated. Subsequently, however, Ham's grandson Kanaan stole Israel from the children of Shem and renamed it. God's promise to Abraham generations later, therefore, was His way of returning what had been taken. Our bond with this land is deeper than any other nation's connection to its own land. A Godly promise, an inheritance for all time, can never be broken or usurped by human beings, no matter how violent or duplicitous they may be. It was by Divine will that a holy kingdom was to be founded in Israel, built by the Jewish people, that will shine its light to the rest of the world. In other words, God's purpose is that the entire world becomes a "holy land," beginning from a specific nation, in a specific place. From there the holiness is to encompass the entire world—as the prophet Isaiah proclaims: *"For from Zion shall go forth the Torah, and the word of the Lord from Jerusalem."* The vision is that Israel is the model of superb, divinely inspired society, marked by morality, well-being, and abundance. The nations of the world will learn from Israel and emulate its ways, on a path toward perfection.

That's why every now and then something happens in Israel that shifts the entire world's attention to the Jews. Israel is such a tiny place it ought to have been completely ignored in the global media, yet it always manages to find itself at the center of attention. God keeps focusing everyone's eyes on Israel, because by putting the spotlight on the Jews and handing them the global microphone, he is empowering them to reveal the divine truth.

Unfortunately, we rarely tell the whole truth, relying instead on "moral" proclamations such as, "We have the right to defend ourselves!" What we really should be saying is, "Hello world, we are the chosen people, and this is our holy land. We will model for you what morality and holiness really mean. Needless to say, we are held accountable for doing so. We have been tasked with becoming the educators of the world—so listen up and start behaving like the model students you were meant to be."

Who Invented "Palestine"?

As mentioned above, "Palestine" was one of those names imposed on Israel by conquerors who tried to do the impossible and remove the holiness from the Holy Land. In fact, history shows that there is no place originally named Palestine, and there never was one. Accordingly, there is no historically recorded national identity group who can accurately be called "Palestinians." Who was the prime minister of "Palestine" before Yasser Arafat? What was the Palestinian flag before the PLO was founded in 1964? What distinguished culture do "Palestinians" have? No one and nothing.

Palestine wasn't actually considered a "modern nation" by anyone until the 1960s, after the Six Day War, when Arafat colluded with the Soviet KGB to make it so. It is a place defined by its resistance to the Godly vision of Jews settling in their homeland. This resistance consists of stealing the Jewish narrative, claiming the Jewish land as their inheritance, and even imitating Jewish victimhood, insisting they are the true victims. The Palestinians present themselves as the natives and portray Palestine as their "promised land" that must be liberated. They portray themselves as the new Jews.

This identity theft goes back a long way. Ishmael was Abraham's first son, his beloved son, and at first his only son (from Hagar). Furthermore, since Ishmael was circumcised when he was 13, he considered this a form of sacrifice that preceded and preempted the "Akeda" of his half-brother Isaac. These virtues made Ishmael feel that he had more rights than Isaac regarding the inheritance of their father's blessings. Sarah saw this identity theft right away and decided to send Ishmael to the desert. She knew that underneath his patronizing moral claims ("I was here first," "I'm a bigger victim," etc.) he was an imposter claiming to be the heir.

Fast-forward a few thousand years and we see virtually the same scenario: those who claim themselves to be the descendants of Ishmael believe the land is their inheritance. Now the question arises: how and why did this negative facsimile of the promised land come to pass? What role does it play in history?

It is a core principle in Torah, and in Chassidic philosophy in particular, that for everything positive in creation there exists an equal and opposite negative force. For instance, because humans have the capacity to feel pleasure from holiness, they can also experience pleasure from unholy things. Whenever something holy is born, something unholy reacts to it. There is truth, and then there's a reaction to the truth: falsehood. The good news is that lies can never express original ideas—all they can do is "suck" their energy from the truth.

Accordingly, when Israel was born, an anti-Israel entity had to have been born with it. And there is a purpose to this "balance" between holiness and unholiness: it is to facilitate a human contribution to God's ultimate plan. Yes, God promised us the land, but that's not enough; now the land must be acquired through *human* effort. What was once Kanaan, and more recently "Palestine," must become the holy land, by virtue of our human devotion to the will of God. In this spiritual sense, to conquer the land means to remove the unholiness from it. And this process needs to be complete, because if there were to remain any trace of unholiness, it might grow stronger and lay claim to the land again.

That is why there have been so many wars in Israel:

the victories have never been complete. By making the land holy *ourselves*, we are helping to make its super-natural properties become natural. We are bringing its miraculous quality into this down-to-earth, everyday world.

God is already Godly. He always has been. So for *Him* to make the land into the Holy Land isn't much of an achievement at all. *Our* mission is to *become* Godly, by devoting ourselves to His eternal will. It's up us to do our job.

What Is the Nature of the Jewish Soul?

Although Jews may look like any other nation in physical appearance, they also carry a spark of the essence of Godliness within them. Unlike the rest of the world's inhabitants, the Jew was not created solely by God's "speaking" him into existence, but rather was an expression of God's innermost essence. The Jewish soul is Godly and eternal, which means it precedes creation itself. This places the Jew, potentially at least, above the laws of nature, and therefore timeless. As God stands above the limitations of the created world, so too do the Jews. They are not a nation *of the world*; there are nations of this world, and then there is the Jewish nation, not of this world.

This "outsider" status of the Jew is a major cause of antisemitism. We are like strangers to this world, and so we are easily accused of stealing and usurping it. All nations are called *Amei Haaretz,* meaning "the nations of the Earth." Having been created in this world, they inherently belong to it. We can see this in the title of that famous Pop song of the mid-1980s, "We Are the World"—a very apt description of the nations of the world. (The song was recorded by a large ensemble of some of the era's best-known singers and musicians, among them Bob Dylan. Perhaps Dylan, who is a Jew, shouldn't be singing it alongside the other, non-Jewish, singers, who *are* "of the world.")

One of the ways Jews are inherently different is that in essence they are not subject to conversion. A Jew can become the Pope and live in the Vatican, and he'll remain Jewish. This means he'll still be obligated to follow all the Jewish laws, and his children (if born to a Jewish mother) will still be Jewish as well. This means there are certain kinds of human experience that are not consistent with our essential characteristics as Jews, and therefore don't pertain to us. Anyone else on earth can choose to be religious or secular, or switch religions, or abandon their previous religious connections. But Jews will always stay Jews.

Many times throughout history Christians have tried to nullify this reality by making it their mission to convert Jews to Christianity, only to discover that this is ultimately impossible. In the Spanish Inquisition, for example, the inquisitors did everything they could, using all conceivable horrific measures and threats to uproot the Jews. Yet they ended up seeing the "converted" Jews secretly lighting Shabbat candles and putting on Tefillin as if nothing had changed. There was nothing left for them to do but burn the Jews at the stake for their stubbornness. But in truth it was not stubbornness. They simply could not stop being who they are.

And just as we say that a Jew cannot become someone other than a Jew, Israel cannot become another country. Why? What's the connection? The answer is that Jews are sent here to complete a mission: to make of this world a dwelling place for God himself. And the Jews are inseparably bonded with Israel, because the ideal, most conducive place to begin to fulfil this mission is the Holy Land. The Godly Jewish soul descends to the world of physical reality for the purpose of elevating it to a higher, metaphysical reality. Therefore Jews are not "*of* the world"—they are sent *to* the world. Christians know this, but they limit this insight to a single Jew. "He is not of this world," their sources state. "He is of God." True enough, but it's not only about him. It is true of all Jews. It is the transcendent soul of the Jew that enables him to turn a natural world into a Godly world. How do the Jewish people respond to this inherent "outsider" status? In our weakness and insecurity we often try to claim that we are indeed of this world, like everyone else. "We're just like you. See how tolerant and democratic we are! Look how many Nobel Prize winners we have!" This argument does not play out very well, because the nations know we are not telling the truth. Moreover, they demand that we demonstrate to them who we really are—to help them recognize the true nature of *their* existence. Much like a teacher who enters a classroom, we should never present ourselves as one of the students. The nations leave us with two options: either teach, or leave. The expression *Ha`am Hanivchar*—the chosen people—is frequently seen as provocative or controversial. What it means is that we are not native to the earth, but to the heavens. Similarly, the phrase "God's children" means that we originate in Godliness. That is why our existence has always appeared miraculous, other-worldly. The birth of Isaac to Abraham and Sarah, for example, was completely miraculous. Sarah was barren at 90, and Abraham was 100 years old, but this did not stop them; which goes to say that right from the start our very lives have been extra-ordinary. And since then we continue not only to thrive against all odds, but to evolve spiritually as Jews. This, despite having been subjected to the most brutal assaults known to mankind.

In the week after the horrible Hamas attack on October 7th, 2023, there were, unbelievably, a number of anti-Israeli protests around the world. In one such incident, on an American University campus, there was a Jewish professor who stepped up and spoke against the protestors' immoral and insensitive behavior. The next day a student came to his office with flowers. She told him how inspired she was by his bravery. "I can really see that you are the Chosen People," she said. The professor, well known as a professed atheist, replied, "You know I don't believe in all that stuff, right?" Her response: "You may not believe it, but everyone knows it's true." When we attempt to deny who we are, we come across as dishonest.

The antisemitic perception of the Jew is that he is sneaky, forever plotting something nefarious, always concealing vital information. "He must be up to something terrible," the antisemite is thinking. "They're trying to take over the world!" People don't like strangers in general. But they especially dislike strangers who refuse to identify themselves. Instead of trying to blend in, a Jew should stand tall and show that he is proud of his Jewishness. This will gain him the respect he deserves.

Who Inherits the Land Given to Abraham?

Two Biblical personalities— *Eisav,* Jacob's twin brother, and *Ishmael,* Isaac's half-brother—have been seen throughout history as archetypical enemies of the Jewish people. Eisav represents Western civilization; Ishmael represents the Arab world. They are both descendants of our patriarch Abraham. Though cut from the same cloth, during the last two millennia they have come to be associated with two very different religions: Christianity and Islam, respectively.

The Torah describes Eisav as a hunter, a man of the field who lives by the sword, a ruthless conqueror. Jews have suffered from Eisav's tyranny through more than 2000 years of exile, first under the Roman Empire, then at the hands of Christianity.

Like the hunter who proudly displays on his wall the head and antlers of the deer he has shot and killed, the Romans arrogantly exhibited their atrocities and conquests for all to see. They engraved, for example, upon the triumphal Arch of Titus in Rome, an image of the golden Menorah they had stolen from Jerusalem. A hunter has no desire to obliterate his prey. He delights in showing off his trophies. In the same way, the Roman Empire strove to magnify its honour and prestige by exposing evidence of the nations they had overpowered and continued to rule.

Likewise, it was never the purpose of the *next* wave of Eisav's offspring to completely annihilate the Jewish people, nor to dispossess them from their homeland. The militant Christian Church relentlessly persecuted, proselytized, humiliated, abused, and murdered Jews over many centuries. To this day there remains a virulent strain of antisemitism in their midst, and conversion of the Jews is often still a goal. Nonetheless in more recent times Christianity has, to a significant extent, come to accept the Jewish people, appreciate their unique role in the world, and respect their unshakeable bond with the land of Israel.

Not so with the descendants of Ishmael. Ishmael is described in the Torah as a barbaric "*pere adam,*" wreaking chaos wherever he turns; someone whose "hand is against everyone, and everyone's hand is against him." (His hand, not his sword.) Regarding their Jewish half-brethren, the Ishmaelites have convinced themselves that they alone are the true heirs to the land of

Israel, and they are compelled to obliterate the Jewish presence in order to stake their claim.

The problem began when Ishmael was a boy. He was the son of Hagar, Sarah's handmaid. When God promised that Sarah would bear to Abraham a son named Isaac, Abraham rejoiced, but he was still concerned: "If only Ishmael will live to be God-fearing, that would suffice!"

So God blessed Ishmael as well. "I will make him fruitful, and extremely numerous. He will father twelve princes, and I will make him a great nation. But I will establish my covenant with Isaac, whom Sarah shall bear."

As the boys grew, Sarah saw, with prophetic vision, that Ishmael's character and wanton behavior rendered him unworthy of sharing an inheritance with Isaac. "Cast him out," she insisted. Abraham hesitated. God said to Abraham, "Do not be perturbed about the boy and the handmaid. Whatever Sarah tells you, listen to her voice. Your true descendants will be through Isaac."

For those who consider themselves Ishmael's descendants, this was a declaration of war. To be so shunned, to be denied their "father's" legacy, was impossible for them to accept. Ishmael having been the older son, as well as having been circumcised according to God's command, makes it all the more outrageous to them. They know that the land can only be inherited, not won. And they are correct: to conquer the holy land by war is not an option, as history has shown. They are left with no choice but to eliminate all Jews from the face of the earth, including the babies, for as long as there are Jewish people in the world their very existence is a competing claim to the land.

The Ishmaelites, therefore, to this day have no interest in "winning a war" or conquering land. They will never accept any peace agreement or negotiated settlement. Treaties, compensations and accords of all sorts will never work for them. Ishmael will forever refuse anything but full ownership over the entire land.

The proof is that whenever Israel and its allies have offered the Arabs recognition, land, money, or even a state alongside a Jewish state, they have always refused. "You can't give me what is legally mine," says the descendant of Ishmael. "The offer itself is an insult." Nor do they feel threatened by military might, because no military can take away an inheritance for which they are willing to die. The only solution, in Ishmael's eyes, is to exterminate the Jews.

Their claim to what they consider an inheritance is absurd, in light of the fact that today's Arabs cannot even be certified as descendants of the Biblical Ishmael. Over the centuries, intermarriage with descendants of Ham and Yaphet has eroded their Semitic identity. Palestinian "Arabs" are actually a mixture of ethnicities without a coherent identity. Regardless of their adherence to some traditions that can be traced to Abraham, such as their monotheism, circumcision, and prayer— of which they should be proud— these connections are purely *ideological*; whereas the Jewish people can attest to a long, largely unbroken chain of a *biological* lineage.

Arabs often try to question and deny the Jewishness of modern-day Jews, arguing that today's Jews are actually White Europeans pretending to be Jews. After the Six-Day War, Saudi Arabian diplomat Jamil Baroody (he was their Representative at the UN) went so far as to declare, "We don't want you in the Middle East! You are Western colonialists, not the real Jews. We can see from your short skirts and American food!" But biological heirs receive their inheritance by virtue of who they are, not what they do. Jews do not inherit Israel because they are religious or keep to certain traditions and values, but because they are Jews. As the Lubavitcher Rebbe said in response to the Saudi Arabian's speech, "We are the children of Abraham, Isaac and Jacob, and that has never changed. Our status does not depend on how we dress or act. It's a fact of life."

Given this perspective, Israeli leaders would do well to reconsider the language they use in both the give-and-take of international statesmanship, and in their public pronouncements. Insisting that they are "the only democracy in the Middle East" comes across as out of touch with the reality of what their Arab enemies demand, for which they are prepared to murder and brutalize others at the cost of their own lives.

Such points may be applicable to ongoing Jewish struggles against the descendants of Eisav. Eisav is a hunter, a conqueror. He's not so concerned about rightful ownership, because he can take whatever he wants by force anyway. Ishmael, however, is deeply threatened by the Jewish return to their homeland. Their singular concern is that the world will accept their claim to the inheritance.

What We Learn from the Palestinians

In 1962, U.S. military intelligence discovered that the Soviet Union had been installing nuclear-armed missiles in Cuba. The American response was so fierce, there was a very real risk of triggering a third world war. To President Kennedy and his National Security establishment it was absolutely imperative: the rockets must go. The Cuban missile crisis demonstrated to the world that America would not tolerate such a threat to its entire civilian population.

After Israel "disengaged" and abandoned Gaza in 2005, the Israel Defence Forces (IDF) were well aware that Hamas, along with its other kindred Palestinian terror organizations, were collecting and concealing an astounding arsenal of weaponry. Rockets, bombs, artillery, guns, knives, drones, grenades, and more were hidden beneath hospitals, schools, mosques, private homes, and underground tunnels. The purpose of these weapons, as they state clearly again and again and have embedded in their mission statements, is to annihilate the Jews. What sane country would allow its mortal enemy to amass and deploy so much ammunition right next door?

Yet Israel has allowed, tolerated, enabled, and even funded (!) the growth of its own enemy's aggression. The

Torah tells us that someone who is overly gentle, merciful, and trustful can actually empower the very evil that threatens his survival. In that sense, the livelihood and legitimacy of the people we call "Palestinians" have drawn their life force from the Jewish people and their innate kindness. There is no Palestine without Israel, and there are no Palestinians without Israelis.

Tanya, the foundational book of Chassidic philosophy, explains further that darkness is not an entity on its own, but merely the absence of light. In the same vein, evil is only as real as we allow it to be. The tendency of good people to avoid conflict, whether through passivity, ignorance, or neglect, can allow evil to prosper and thrive. The gardener whose garden is overrun with weeds never intended to let them grow. He just wasn't paying attention. The Palestinians are very much the same. If we know as a historical fact that they never existed as a distinct nation until they started resisting Israel, why do we even refer to them as "Palestinians"? God gave us the land (see Rashi's commentary on the very first verse in Genesis.) Why then do we keep offering them portions of it from which to attack us? We are empowering them—creating them!—with our fear.

This is the Torah's message to the Israelite warrior: "When you go out to war against your enemies... have no fear of them. For your God, who brought you from the land of Egypt, is with you." The secret meaning of this passage is in the exact Hebrew phrasing: "When you go out to war *above* your enemies." When the enemy attacks, the key to victory lies not in confronting him on his own terms, but in *rising above* the fray with the knowledge that God is with us and fighting for us. An even deeper way of understanding this lies in the fact that concealed within the enemy is a divine spark, whose purpose is not to intimidate or frighten us, but to *elevate us.* The hidden role of the Palestinians in this conflict is to lift us high above our own sense of powerlessness, stop feeding the enemy with our fear, and courageously embody the triumphant power of God.

The way the Palestinians sustain their existence is through hatred, usurping our Jewish energy through unfathomable violence. They sense that if they weren't so horrifically brutal we would never even give their envy a second thought, let alone negotiate with them. We only take them seriously when they pose a threat to our security. Consider the fact that we rarely feel the need to discuss in depth our relationships with other minorities in the region, such as the Druze or the Bedouins, because they are not violent toward us. But the Palestinians are dangerous, so we give them an identity. We recognize their demands, expressed in terms of land, money, and submission to their terror. Even when we erect expensive security fences and invent high-tech Iron Domes, it's tantamount to saying, "Palestinians are real, and they're not going anywhere. We have no choice but to live with this problem forever."This is not how a strong, mature state behaves. A confident, self-respecting state shouldn't have "enemies." When a child is misbehaving, he is not an enemy — he simply needs discipline. The father must take charge and reveal his inner strength by expressing authority and wisdom.

The purpose of our taking back the land that we so foolishly surrendered, therefore, is to make God's will and presence real. It's for that reason that God gave the land to the Jews—to Abraham and his children and children's children—and we must insist that His will is obeyed and His truth revealed.

On a still deeper level, when the nations claim that we do not have a right to "the land," they do not mean only Israel. They mean the earth itself. "If Jews are Godly," the world protests, "they belong in heaven, and they cannot have dominion over the earth as well." To the casual observer that might seem to be a valid point, if not for the fact that God assigned us the task of coming down into this physical realm, to conquer the earth by making it Godly. The rules of the game have changed. The limitations of human logic no longer apply. We come from a higher source.

Suppose the Jewish people were to say, *"Yes, we understand: God gave us the land and appointed us to conquer it in His name. But practically speaking, politically, we just can't get away with it."* Then we might as well be worshipping idols. To imagine that God is competing against the "international community" or the world's politically correct opinions, or any other "pragmatic" quicksand, is a form of idolatry. It is like saying, *"If we lived in Heaven, we could have done many heavenly things. But since we live on earth, we must heed the earthly rules."* It's as though we're making the creation more real, more relevant, and more powerful than the Creator Himself. God does not want to remain in the realms of religion and faith only, but to express Himself in all aspects of life. The mission is to make God so real, so visible, that the material world itself surrenders to His light, and falsehood surrenders to truth.

Political Babble and Jewish Statehood

Israel's Struggle for Autonomy

There is a major difference between a country and a sovereign state. A country is a tract of land with defined borders, which a distinct ethnic or national group calls home. It might enjoy a certain measure of autonomy, but it can also exist under the rulership of an empire, or some other colonizing force. The term sovereignty, on the other hand, describes a completely independent, self-governing state. If no other country can tell your country how to govern itself, if your country is free to pass laws and pursue its own policies without foreign interference, and if no laws are imposed upon your country by any external power, then you are living in a sovereign state.

A country's sovereignty can be undermined from the outside by a more dominant and influential force, whether political, fiscal, or military. This dependency creates what social scientists call a "banana republic," where the local leaders are simply puppets of an empire. The country itself is merely a tool in a larger, perhaps global, economic and political scheme. A country's sovereignty can also be undermined from within, by citizens who believe that the current rulership is illegitimate.

Sadly, the State of Israel is not yet sovereign. It has become financially, geo-politically, and ideologically dependent on America. Even within Israel's own citizenry there is an extremely hostile minority that does not fully recognize Israel's right to exist. Some seem more committed to protecting the enemy's civilians than their own.

Does this truly serve our security and statehood? Or are we trying to be accepted by the Western world? Does it represent *our* values and faith, aligned with our tradition, or is it a reflection of American or European values? Obviously, it is not simple for a small, young country to release itself from the chains of its sponsors. It requires a great deal of courage and strength to face up to such powerful allies. But let's now ask ourselves: does this relationship benefit us? Will it bring us to the fulfillment of our destiny?

An autonomous nation makes its own rules, without being intimidated into making decisions that neither benefit its people nor honor its culture. Take the government of Singapore, for example, where anyone caught chewing gum in the streets can be fined up to $100,000 and face up to two years of imprisonment. Or Dubai, which has only one prison, because most of its criminals are either deported or executed. Are these rules "normal" through Western eyes? Obviously not. But this is true sovereignty.

The truth is, Israel cannot achieve true sovereignty without the coming of Moshiach—the ultimate King of Israel, the divinely empowered Jewish leader who imbues Israel with strength, and the entire world with justice and mercy. Without Moshiach, Israel remains a stepchild of America (and is deeply hated for it). Freeing ourselves from external oppression starts with using our ability to defend ourselves. Saving Jewish lives is a mitzvah of the first order. Just as an Arab terrorist murders Jews because "the Imam told me to," an Israeli leader will protect the Jews, proclaiming, "Our rabbis told me to. Our Torah tells us to." That is where Sovereignty will begin.

The Trap of Democracy

Simply put, sovereignty basically means, "We do what is best for us." Seems like a good idea, especially in dangerous times such as these. Why then is it not yet happening in Israel?

One reason is the fact that Israel is a democracy. That's not to say that democracy is a bad thing. But is it working for us? Do we truly think it's so wonderful being "the only democracy in the Middle East"? Why are we so proud of that? Here again we must ask: are we adopting this designation because we're seeking approval and legitimacy?

Democracy is a very good system of government, but it may not be the best one in the Middle East (which could explain why we are the only ones practicing it). A Euro-American democratic government, founded on democratic principles but surrounded by ethno-religious dictatorships, will naturally be very vulnerable, and will always be seen as an outcast. In that sense democracy may be hurting Israel badly, because democracy doesn't belong in the ecosystem in which it must survive.

Also—a Jewish state can never really be fully democratic, because the system itself contradicts our own ethno-religious foundation. Technically, our enemies could gain a majority in the government and vote to kill us, and it would be a "democratic" genocide. So it's crucial that we define Israel first and foremost as a Jewish state, meant for the Jewish people. Then we can add that this state uses democratic values to govern itself— but only when these values are beneficial to our wellbeing. This pragmatic approach to democracy is the only way to survive in a region full of hostile dictatorships.

Generally speaking, democracy reflects the values and beliefs of the people. That is why it can be a very corrupt and chaotic system if its people have the wrong values. Democracy may work for people who are already moral, but it won't *make* anyone moral. Yes, the majority can choose its leaders and be run by majority rule, but who can guarantee that this majority will be honest, moral, ethical, and Godly?

Torah principles are essential to a democracy, ensuring that its citizens will select its leaders and laws on the basis of a clear moral compass. A society that studies and understands the message and the meaning of Torah will know right from wrong, holy from unholy, and wholesome from destructive. In other words, democracy is a good idea, but only if it follows the Ten Commandments. Once a society has a sound moral infrastructure, its values can be expressed and applied magnificently through the governing tools of democracy.

For example, the original vision of American society included the idea of religious freedom: that nothing and no one can define or impose upon another person's relationship with God. It was founded by people who had escaped religious oppression in Europe and wished to build a New World, where every human being can freely develop a personal approach to Godliness. The Bible played a major role in this vision, and as a result this democratic "experiment" was a great success. Slogans such as "In God We Trust" became the bedrock of American civil society. Family values, patriotic sentiment, and global policies were made in the image of divinity.

In recent decades, American democracy has become dysfunctional. Its culture has become torn, confused, and angry. It began when the Ten Commandments were erased from the public sphere. Godliness is ridiculed.

Exposure to religious thought and the freedom to practice school prayer have become a thing of the past. The result is a democracy without values. Real freedom of speech, respectful inter-group relations, traditional views of gender, classical education, a stable sense of security—none of these are functioning properly today in America. Similarly, when Israel takes pride in being the only democracy in the Middle East, it mostly refers to its free elections. Open elections are great, but you must be well educated to know whom to choose. Recently, Israelis have struggled as never before to form a government, largely because the politicians do not reflect traditional Jewish values anymore. Empty Western slogans inspired by Hollywood just cannot inspire the people of Israel, and this cultural void creates chaos.

Another democratic principle in which Israel takes great pride is fully accepting anyone and anything according to the ideologies of liberalism and pluralism. However these values have ironically been emptied out of other values, because so many diverse opinions and perspectives are now considered "racist," or otherwise intolerant. The attitude of "live and let live" has framed all other diverse values as "undemocratic." Democracy has become allergic to religious values. Only a Godly approach to democracy, or a democratic approach to Godliness, will restore the integrity of the political system in Israel. Otherwise it will continue to sabotage its own spiritual development.

The Immorality of Equality

There are certain words we tend to use without bothering to define them accurately. They're in common usage, and therefore obvious: it's easy to assume that everyone agrees as to what they mean. One of these is the word *equality*. Clearly a good thing, right? Only a monster would oppose the idea of equality.

But what exactly does equality mean? Is it always, unquestionably, a virtue? Maybe we should not be in such a rush to put the concept of equality on a pedestal. Is it always a good thing? Are we all equally tall? Equally smart? Equally talented? Equally deserving? Are all such expressions of equality a good idea?

In these modern times, *equality* has come to be considered more or less equivalent to *morality*. Conversely, inequality is seen as amoral. But they are two different things. Morality means, "If I am stronger than you, I'm still not allowed to oppress you." Or "If I am richer than you, I cannot exploit you." Or "If I am older than you, I will not take advantage of your youth." The whole idea of morality, therefore, is that the *big guy* should not beat up the *little guy*.

A moral government doesn't take away the freedom of the individual, though it certainly can. The government has the money, the police force, and the power to enact legislation that could completely take away individual rights. Similarly, the idea of animal rights means we are not allowed to abuse them, though we're so much stronger and/or smarter than them. So morality really means refraining from being cruel to others, even when you have the power to do so. And the reason why you can be cruel is precisely because there is no equality. The basis of morality, therefore, is the recognition that we are *not* equal.

Due to humanity's long history of cruelty, there's a very prevalent belief that power (i.e., inequality) inherently corrupts, and therefore no one will ever truly practice morality. The remedy, according to this point of view, is to affirm that "we are all equal" (or that at least we should be). For instance, cruelty to animals is prohibited nowadays because an animal is presumably equal to a person. Which is of course untrue, a figment of the imagination, or at best wishful thinking. In other words, a lie.

Maintaining such a lie is not easy. The proponents of the lie have to be pretty aggressive to put the lie across consistently. A fundamental principle of socialism, for example, is that since people are inherently corrupt, we must *impose* equality on them. "To prevent the greedy from stealing," a socialist might propose, "we need to take away everything that can be stolen." This wouldn't make us moral, of course; it'd just remove the possibility of behaving cruelly. In the absence of morality, it seems as though only absolute power can stop people from acting out their corrupt nature.

This idea of equality as a universal humanist principle is a product of modern democracy. One of the pillars of democracy is that when it comes to political rights and the ability to vote, all citizens are presumed equal.

A democratic government is not allowed to discriminate between its citizens, even though they are indeed different. The concept of equality, therefore, was never meant to be seen as an essential, natural belief that all people are equal. That is just empirically wrong. It applies only in the eyes of the law.

From the divine perspective, of course, all humans are descendants of Adam, and are therefore equal in the sense that we are all created in the image of God *("b'tzelem Elokim")*. But that doesn't mean we are equal in all aspects of life. God provides life to his creations on two levels, simultaneously—collectively, and individually. And the way we can come to understand and sort out this apparent paradox is through the core principles of Torah.

The first level is relevant to all creations equally. By the mere fact that God is the source of everything that exists, flowers, animals, and of course all human beings are indeed equal. On the other hand, however, God invests each individual creation with specific, unique characteristics, traits, abilities, and roles. In that sense, the creations are not equal at all, but rather exist along a spectrum of different aptitudes and purposes. So both viewpoints are true: we are all equal, in general, while at the same time we are each singular, unique, and often find ourselves in hierarchical relationships.

It's when we confuse the two perspectives that we get into trouble. If we insist on our "equality" when we ought to be appreciating our differences, or if we weaponize our differences when we should be living in harmony as one, we become dysfunctional. Whether between individuals, between species, or between nations, such fundamental misunderstandings can easily escalate (if not inevitably explode) into war.

The Naivete of Compromise

There's a familiar Biblical story (1 Kings 3:16) about the wisdom of King Solomon, and how he adjudicated a dispute between two women who both claimed to be the mother of the same child. Solomon ordered that the baby be cut in two, with each woman to receive one half. The first woman accepted the compromise—half a baby, after all, is better than none. The second, however, hoping to save the child's life, asked Solomon to give the baby to her rival. In response, King Solomon ruled in favor of the second woman. He knew she was the real mother, because the life of the baby was so much more important to her than her own happiness.

Since 1917 the world has been offering a two-state solution to the Jewish and Arab populations, in the region of the ancient Land of Israel *(Eretz Yisrael)*. The Jews have consistently accepted these proposals; in 1948 this resulted in the establishment of the State of Israel. The Arabs, however, have always rejected a compromise, insisting that they will not settle for anything less than the entire land of what has come to be called "Palestine." Yet the issue persists. Many Jewish citizens of Israel are actually still in favor of the two-state solution, while the majority of the Arabs still refuse.

In light of Solomon's commanding sense of compassion, who should be considered the real owner of Israel? In our times, kindness and mercy are seen not as signs of great virtue, but as weakness. When it comes to this seemingly never-ending conflict over the Holy Land, whenever we have given away parts of Israel it has only strengthened the enemy's stubborn claim: that they are the rightful and exclusive owners of the land, in its entirety.

The controversial Rabbi Meir Kahane once said, "I can really understand the Arabs. They want this land, and so do I." Yet he recognized that this land is ours not simply due to our own desire, but because it is unequivocally designated as such in the Torah. It is the will of God, and therefore our abiding passion, to bring God's promise into being.

In its confusion about this conflict, the West can only offer shallow, childish solutions. "Why can't you just share? Everyone must compromise. Each one should be satisfied with his half." These so-called answers to the impasse are based on the premise that there is no truth and no justice—just two sides arguing over nothing. How patronising! Particularly in view of the fact that we know with certainty, through hard-won experience, that a two-state solution will not make life in Israel any safer. Moreover, Muslims consider it an affront to their claim that the land is already theirs by inheritance. As such, the "two state solution" has only prepared the ground for a much more horrible war.

Imagine someone setting foot in the house you've built with your own two hands, saying, "Let's split this place into two homes!" Would you even consider something so absurd? It's ludicrous even to speak about a "peace process" when the hatred and the enmity are so loud and clear. The simple fact on the ground is that granting statehood to a mortal enemy will not bring peace. Perhaps, ironically, we are fortunate that it will never work—because its very impossibility will encourage us to find a real and lasting solution, one that reflects the timeless Jewish values of justice and truth.

Bringing Light to the UN's Darkness

The United Nations presents itself as an organization dedicated to maintaining international peace, developing friendly relations among nations, and serving as a center for harmonizing the behavior of member nations all around the world. Noble goals, one might think, but very far from the truth. Throughout the years we have seen this organization turn into an incubator for antisemitism, where cruel dictatorships join hypocritical democracies in delegitimizing the Jews, under the transparent double standards of liberal and humanistic slogans.

Brutal terrorist attacks on Israel, such as the horrific

events of October 7th, 2023, do not change the UN's tone one iota, making it even clearer that this is a corrupted institution based on nothing but lies. Not a single word of truth is heard; we already know what anyone who stands up to speak will say. It's all about global dominance and personal gain. Countries who are at each other's throats outside the UN gather together and pretend to be friends, just to demonize and dismantle Israel. It's a mystery how anyone can still acknowledge this place as a legitimate institution. What is most scary about the UN is that it represents the state of the nations in microcosm. As a platform for fomenting hatred toward Israel, it is emblematic of all the falsehood in the world.

At the UN's 60th anniversary, former Israeli Prime Minister Ariel Sharon came to give a speech. He spoke like a wise grandfather speaking to his grandchildren. "I arrive to you from Jerusalem, Israel's eternal capital for over 3000 years," he said. "I stand before you at the gate of nations, as a Jew, a representative of an ancient people." Unabashedly expressing his Jewish pride, he went on to say, "The Jewish people have a long memory, originating with our father Abraham, and continuing with the giving of the Torah on Mount Sinai."

The substantive remainder of his speech may be debated, but this was a moment from which we should all take inspiration. Sharon restored the relationship between the Jews and the nations to its natural order: we are the "grandfather" of the world. If proud Jews were the teachers, and the world, the classroom, the

UN would be a perfect platform for disseminating messages of holiness, truth, and justice. But when we act as though we are the students, it's a disaster waiting to happen, because no one gives us any respect.

So let's stop lying to them, and to ourselves, and tell the world who we really are. We are the most ancient people to have witnessed the most glorious and the most terrifying events in history, and survived. We have stood as a nation, received Godly revelation, and preserved it in its original form down to the most minute detail. We know how the world ought to be behaving, and where it is headed instead.

When Binyamin Netanyahu became Israel's Ambassador to the UN, the Lubavitcher Rebbe said to him, "The UN is a place of darkness. Your task is to bring some light."

The Jewish Ethics of War

If God is good, kind, and merciful, why did He create war? How could bloodshed, death, destruction, and suffering—call it the *hell of war*—be considered justified, let alone necessary? Can there be such an oxymoron as a holy war? We've spoken earlier about the kabbalistic idea that everything good in creation is mirrored by an equal and opposite evil twin. But to suggest that the sublime human ideals of peace, love, and brotherhood cannot exist without hostility, hate, and conflagration... seriously?

That war exists at all does indeed mean that it must serve some purpose in God's plan. Certainly there have been throughout history pointless, unwarranted, even unforgivable wars, driven solely by the lust for murder or theft or subjugation, and sometimes masquerading as holy wars. But the purpose of an actual holy war is ultimately to restore and amplify Godliness in the world.

This idea can be better understood in terms of the inner conflict that takes place within each of us. When we are faced with a choice between our human side and our Godly side, an internal war breaks out. The very fact that this battle between right and wrong calls for exercising and refining our freedom of choice renders it, in a very real sense, a holy war. Whenever we have to make a choice, we are up against this test, and it's not always easy. When we choose God over something unGodly but (for the moment at least) equally tempting, we are strengthening our bond with Him.

There are two types of evil that we must overcome in our personal wars, though both can drive us to sin and defy God's will. One type is the pure evil that cannot become refined or transformed, but must be utterly eliminated. The other type is an expression of our natural self-centeredness as human beings. It should not be destroyed, because it can actually provide us with vitality and promote a positive, generous sense of self. Its selfish nature can be, and should be, rectified, by making it Godly. For example, upon recognizing one's own selfish urges one can "turn them around" and choose to be more compassionate. Through a process of education and discipline, our human energies can be brought under control and directed toward good deeds. In other words, there is irreversible darkness, and there is darkness that can and must be transformed into light.

This process is not confined to our inner wars. It is also expressed in the historical wars between nations. In wartime, even in a just war, people can be very perplexed about the purpose of the war, the type of evil they're up against, and the role of God in managing the conflict. In Israel, for example, we see empirically how politically motivated division and international pressure can alter the course of a war, leaving Israel insecure, hesitant, and toxically apologetic. Such confusion can cloud the minds and hearts of generals and statesmen alike.

Among the most misunderstood issues capturing our attention in this war-torn world are *civilian casualties, proportionate response, and hostages.* Let's shed some light on each of these.

Protecting Civilians Responsibly

Hamas terrorists have declared that they will never consider Israelis as civilians, because they are all, indirectly if not directly, supporting the Zionist cause. Their definition of military personnel, therefore, includes Jewish men, women, and children, from infants to elders. None are innocent. How then should we identify *their* civilians? If we were to dismiss the very presence of "innocent civilians" we would be adopting the same policy as Hamas. But we are obliged as a people who represents goodness and morality to follow the ethics of the Torah. In the Torah we learn that murder is forbidden, and that the killing of innocent civilians during wartime is not allowed. In that sense, we are categorically and ideologically different from them. Should that harm our chances of winning the war?

Before we tie our hands behind our backs on the

battlefield, we must ask: who is a civilian? If a man knowingly works at a factory that makes bombs meant to kill Jews, is he an innocent civilian? If a woman cooks food for the terrorists so they can go and kidnap Israelis, is she an innocent civilian? Furthermore, if a woman makes babies with the purpose of raising terrorists who will "Free Palestine" from the Jews—especially if those children are then indoctrinated to want to murder Jews—is she "just a civilian"? No, they are all part of the enemy's fighting forces. We cannot draw an artificial line separating between terrorists and the society that nurtures them. To paraphrase an old saying, "It takes a village to raise a terrorist."

Is the average Palestinian in Gaza not a threat to every Jew? If he votes for Hamas, cheers them on, and celebrates their terrorist attacks, is he innocent? In an interview with the son of a Hamas leader who managed to escape Gaza, a reporter asked him if Palestinians generally support Hamas. He said, "I promise you, when Hamas is defeated the people of Gaza will dance in the streets!" Does that mean they are "innocent civilians"? Not necessarily. They will more than likely be happy because they have been released from an oppressive dictatorship, not because they have renounced violence toward the Jews.

So how can we know who is an innocent civilian? Very simply: an innocent civilian is a person who surrenders. If he says, "I give up," "I am not part of this war," "I am not here to fight you," "I am waving a white flag," "Please help me"—we should not kill him.

The most important point here is that it is *not* Israel's responsibility to locate and identify these people. The fact that someone is not wearing a uniform does not make him an "innocent civilian." It is up to him to make himself known and visible as someone who is surrendering. Otherwise our soldiers will be distracted, our military operations restricted, and our ability to win quickly and efficiently severely undermined.

It is a very good policy to do all we can to avoid killing innocent civilians, but it is ridiculous and immoral to assign this mission to soldiers on the battlefield. It is not the responsibility of Israel to distinguish between those who attack, those who support the attack, and those who object. It's up to *you* to show *me* that you're not here to fight me. Referring back to our point on the two types of evil: the hostile Palestinian who wants to murder Jews belongs in the category of utter evil, and must be destroyed. A civilian who surrenders, however, is in the category of redeemable evil and should be properly educated. That's the sensible, ethical Torah approach to civilians during war.

Can Proportionality End Wars?

Whenever Hamas bombards Israel with rockets, as they did during and after the attack on October 7th, Israel responds by bombing terrorist enclaves and buildings in Gaza. It's at this point that the global media and activists begin protesting that although Israel "may" have a right to defend itself, the response should be "proportionate." In other words, when the IDF attacks it should not be harsher than the attack to which it responds. Is this a Jewish idea? Does Torah tell us to fight proportionately?

Yes. The Torah guides us to remain proportionate. But what precisely does that mean? If you kill ten of my people, I should kill ten of yours? You rape three of my women, so I rape three of yours? This approach is unbelievably infantile and immoral. *Proportional* means that my side must eliminate as many opponents as needed to establish peace. If I can do so by killing one person, then I am not allowed to kill more. But if it is necessary to neutralize a million people to achieve peace, that is what we must do. Proportionality, therefore, is about doing what is necessary to end the war. Not more and not less.

The logic behind this policy is that a premature ceasefire that fails to put an end to the conflict is profoundly disrespectful to the casualties thus far, because it means everyone who died has died in vain. If the fighting is "paused" in the name of humanitarian proportionality, only to inevitably begin again after the enemy regroups and resupplies, all the previous deaths become pointless. This is totally immoral toward the casualties on both sides.

A surgeon cannot stop in the middle of a critical surgery because the patient is in pain. The only way to "justify" someone's death during a war is if it contributes to the end of the war. A holy war must continue until the problem is eliminated once and for all.

The term "genocide" is often used in the media to label situations in which a great number of civilians have died. The purpose of the term is to cast the killing as a crime against humanity, as a ruthless attempt to wipe out an entire group. But the word *genocide* applies to killing people who are *not* a threat to your life. If the people killed have surrendered and have no murderous intent, if the purpose is to erase a specific ethnic group from the face of the earth, *that* is genocide.

But if the only possible way to achieve peace and disable an insatiable enemy is by killing all those who take their side, this is a life-saving procedure, not genocide. Doing whatever it takes to bring peace and ensure that the next generation will not have to wage endless war is true morality.

What if we're pitted against a people who do not care about death? How can we end a war with those who see death not as a tragedy but as religious virtue? In that case we must find another way to end the war. It must be something that the enemy cares about so deeply, the threat of losing it will make him stop. The moral way of fighting a war is to accomplish peace as quickly as possible. It's not peace unless it guarantees that there will be no threat of hostility for at least 40 years.

Are We Encouraging Kidnapping?

According to Torah law, if a person is kidnapped, his community must redeem him. The freeing of a captive is a great mitzvah. But by the same token, Jewish law does not allow the redemption of a hostage if the price is too high. It will encourage the kidnappers to turn one kidnap victim into an industry. Having ransomed one captive and brought him home will make everyone a target of kidnapping in the not-too-distant future.

For decades, Arab terrorists have been kidnapping Israeli soldiers and civilians for the purpose of blackmailing, demoralizing, and destabilizing Israeli society. It's commonly known among Arabs that Jews have a special sensitivity when hostages are taken—that the Jewish heart will not allow leaving any Jew behind. The solidarity and sense of collective responsibility among Jews is so great, they will give up almost anything to bring them home. Every Arab knows that one kidnapped Israeli can lead to the release of a thousand murderous terrorists in trade.

It brings us to a deep dilemma. On the one hand it shows how good we can be as a people, and how close we can feel to one another. Yet it makes us weak in the eyes of the enemy. We must look at the issue of hostages from a new perspective, to clear the fog brought on by secular morality.

Consider this: during wars between two Arab states or groups, we don't tend to see Arabs taking hostages from their Arab enemies—because Arabs know there's no point. No Arab government promises to pay the price for saving them. The deeply ingrained Jewish commitment, however, to rescuing our captured brothers and sisters, makes kidnapping Jews a very profitable business. In short, it makes every Jew in the world a potential hostage. So where is the morality?

Similarly, the policy of risking our own soldiers' safety for the sake of "non-combatant" enemy civilians makes every civilian in Gaza into an ideal human shield. If you tell your enemy that he has power over you, that you are willing to do whatever he asks, he knows your weak spots. This is in itself immoral. Saying *we will never shoot children* does not protect them. On the contrary, the terrorists will now use children as shields in whatever ways they can. Just as no Arab country takes hostages from an Arab enemy, they also will not use humans as shields. Here again we see how sometimes our kindness and goodness, our idealism and love of life, far from serving God's will can become the fuel that inflames and empowers evil.

When we think of innocent Jews trapped in Gaza, not knowing whether they will ever come home, the heart breaks. Moreover, if we allow ourselves to hesitate in our rescue efforts, every civilian in Israel—the entire *nation* of Israel—is held hostage by Hamas. We can't really complain that they use hostages as a shield when we keep assuring them that it will work to their advantage. Only when the terrorists are convinced that kidnapping Jews is as useless to them as kidnapping their fellow Arabs, can we hope to see the miraculous return of all hostages, safe and sound.

Peace: First Respect, Then Friendship

Picture an idyllic land where beautiful people of all ethnicities coexist in harmony, friendship, and good will. We all want peace, and though each of us may have a different image of what peace means, it is the highest and most deeply held desire of human beings everywhere. Israeli society educates its children to dream about peace incessantly, and its culture is rich with songs and stories calling for a world without hatred or envy or war. Peace has become Israel's highest aspiration, envisioned as the best possible future for all.

However let's not confuse peace with friendship. Good friends do not need to strive for peace—they are already friends! Only enemies need to work toward peace from a place of enmity and distrust. In this sense, peace means something much more elementary than a vision of children laughing and grownups dancing together in green pastures, smiling ear to ear. Peace means establishing a transactional world where your enemy cannot afford to start a war.

As Jews, having lived in exile for thousands of years, we are desperate for acceptance and recognition. Sometimes we want to be loved more than we want to live. So much so, that we are willing to put our own safety and security at risk so that the nations of the world will hold us in genuinely high esteem. Sadly, this doesn't work. The harder we try, the more such efforts arouse disrespect and rage toward the Jews. Why? Because such a peace is a fantasy. You cannot talk about love and friendship with people who want only to wipe you off the face of the earth. If someone is coming to kill you and you hold out your hand in a friendly gesture, you're out of touch with reality. If *"Let's not kill each other, let's love each other!"* is your motto, you're missing a few steps along the high road to heaven.

First and foremost, we must demand of our enemies that they put down their guns. What will make them do so? Not love. This is not a Disney movie. Peace only comes when the enemy is too afraid to kill you, understands that he simply cannot kill you, or realizes that if he tries to kill you the price will be too high. When your enemy is so intimidated by your presence in his world that he can't afford to make a single move aimed at harming you, that's when you'll have peace. This requires zero tolerance of violence. Even throwing a stone at a Jew should bring such harsh consequences, the would-be stone thrower just gives up on the idea

The only reason America was able to make peace with Germany and Japan during the Second World War is because the Allies had attacked them so fiercely, they are still to this day afraid of having to go through such an experience again. That war ended when so many German and Japanese civilians had been killed, they couldn't afford to fight anymore. That sort of peace is not the most thrilling depiction of a world we want to live in, but it's a necessary place to begin. Peace does not mean love; it's a means of beginning to move from peace toward love.

Once you have established peace, if you also want

love, you must show the world a way to live that will inspire them. If you create a very impressive and inspiring life for you and yours, the nations of the world will begin to admire you, and wish to live the way you live. First you left them no choice but to surrender to your superior power. Now you can bring them to love you through inspiration.

Any experienced classroom teacher will tell you that the first thing you need to establish on day one is discipline. Only then—if you are a good person as well as a good teacher—the students will also come to love you. Discipline is not a harsh word. It's neither nasty nor distasteful to hold a student's metaphorical feet to the fire. It's just necessary. Remember: respect is gained through power; love is gained through positive influence. Peace comes when life is liveable; love comes when life is meaningful. When we finally reach the prophesied stage of "*Out of Zion shall go forth Torah Law,*" the nations of the world will indeed love us for who we are and what we have to offer.

Your Role as a Jew in Dangerous Times

The Antidote to Antisemitism

Shockingly, instead of arousing sympathy for the Jews, some of the most horrific pogroms and massacres inflicted on the Jewish people have triggered increasing waves of antisemitism all over the world. Anti-Israel protests call for boycotts and attacks on Jews on the streets of major cities, on university campuses, in governments, in social media. It creates a frightening atmosphere for Jewish men, women, and children both in and outside of Israel. In many Jewish communities people have taken steps to hide or remove visible signs of Jewishness, such as Star of David symbols on clothing and jewelry. People have even taken mezuzahs down from their doors to avoid dangerous confrontations. But are we really safer hiding our Jewish identities?

It's a phenomenon that spans millennia, and centuries, and decades. In Nazi Germany the Jews were very assimilated into the dominant society. Most Jews looked and behaved like Germans, had German in-laws, spoke German, and took on (at least outwardly) German values and ideologies. This did not help them at all. When Hitler decided that all Jews must be expelled and murdered, he did not differentiate in any way between observant and non-observant Jews. For him a Jew was a Jew and will always remain one. The notion that blending into the general society is a safe and responsible strategy has been proven historically incorrect. Taking down a Mezuzah because of antisemitism is like saying, "I live in a bad neighborhood, so I'll take the locks off the doors." Jewish practices are more than mere symbols; they have the power to protect us, spiritually and physically, and have indeed protected us for thousands of years. The very fact that we are still here after having been persecuted mercilessly through the millennia bears testimony.

If worldwide hatred, ignorance, and terrorism were to succeed in making us ashamed of who we are, it would mean the triumph of evil over goodness. We cannot accept this. The response should be not less, but more Jewish pride, more performance of mitzvot, and greater visibility as the unique nation we are meant to be. This will serve to inspire the world around us and empower our Jewish brothers and sisters in unimaginable ways.

Think of an Israeli soldier. At first glance he looks like any other soldier anywhere. He wears a khaki uniform, carries a weapon, and dons camouflage and body armor. But he also wears tzitzit, puts on tefillin, and cries out *"Shema Yisroel"* as he heads into battle. These uniquely Jewish symbols and practices are seen by IDF soldiers as protecting, unifying, and lifting the spirit of all Jews across the globe, while granting them the courage and confidence to be victors, not victims.

In many ways, every Jew is a soldier. So let's be proud, and protect ourselves, both physically and spiritually, by being more Jewish than we ever have been. We have nothing to feel guilty about, nothing to be ashamed of, and nothing to fear. We have been chosen, not for special privilege, but for the hard-won *merit* of serving God's highest purpose in the world. We love one another, we love our families, we love the goodness within humankind, we love giving, and we love life. That, improbable though it may seem, is why our enemies are so jealous of us. There will come a time when they will leave rivalry behind and recognize—and emulate— our devotion to living purposeful, meaningful lives.

Celebrating Our Identity Ensures Our Survival

When Jews have been murdered in terrorist attacks, they have died because they were Jews. When murdered in the land of Israel because they are Jews living in their homeland, it is all the more tragic—yet all the more significant. They have died not as victims, but as heroes standing firm in the service of a vital cause. The soldiers of the IDF, too, are fighting for an essential cause. The right of the Jewish people to live freely in their ancestral homeland is an eternal right, bound up with our history, and with our destiny.

Every Jew throughout the world should be driven by the same sense of purpose, each employing his or her unique skills and talents to defend, preserve, and improve the lives of Jews everywhere. This does not necessarily entail defending Jewish life like bodyguards, but rather to do our part in protecting the very essence of Jewishness itself. Survival alone is not sufficient to fulfill our purpose. We are commanded to celebrate and fully express our identity as Jews, in body and in soul.

By being more visibly and publicly Jewish, we defy the world's attempt to shame and weaken our people. Not solely by arranging huge rallies and hanging flags—as all oppressed people are entitled to do—but by seeking out and presenting to the world that which is distinctively ours. Being a proud Jew in one's heart is well and good, but being a Jew who demonstrates his pride in action is effective in another dimension. Do not underestimate the power of lighting Shabbat candles, putting on Tefillin, blowing the Shofar, kindling a menorah, learning Torah – all these gifts, given exclusively to the Jews, are also weapons of holy war.

Every Jew Is a Hero

Whenever something horrific happens to the Jews, suddenly voices are raised asserting that they deserved it, and how it must a punishment, and that they must repent. It's as though G-d is waiting with bated breath to punish us and take revenge. It's giving G-d himself a bad name.

We are all far from who we were when we pledged to fulfill the original mandate on Mount Sinai. Although after so many traumas and so much pain over thousands of years, we have often grown numb, we are always trying to do our best with the tools we have. We have been promised God's forgiveness. At this culminating stage of history we should expect only His kindness, mercy, and love.

Each of us, whether a rabbi, a teacher, an influencer, or a quiet member of a community, must do our best to emulate God in giving another Jew the benefit of the doubt. See other Jews as worthy, beyond blame. See ourselves in solidarity, expanding the love amongst ourselves, as one people with one heart, making the world a better place.

Our empathy can extend toward God himself. If you were to encounter a parent who has lost a child, God forbid, you wouldn't try to offer comfort by saying, *"It's alright, he wasn't such a good kid after all. Maybe he deserved to die."* You'd try to show how you too feel the pain. G-d, too, mourns the loss of his children. Don't try to justify tragedy by insinuating, even in thought, that the Jews who have died had sinned. G-d is our Father. His pain is a Father's pain, but to an infinitely greater degree.

The Lubavitcher Rebbe said, regarding the holocaust, that "Even Satan himself would not dare suggest that this is an appropriate punishment for whatever sins the Jews may have committed." Some tragedies are beyond comprehension. We shouldn't try to explain them away. Judaism is not a cold system of reward and punishment. Yes, consequences for misbehavior are an integral aspect of our quest for morality, but they do not define us.

After the sin of the Golden Calf, G-d was very upset with the Jews. Understandably; yet Moshe responded by asking God, "Why are You angry at Your people?" You are G-d Almighty, infinitely great. And now You're allowing Yourself to get bent out of shape over a cow made of melted metal???

We can learn from Moshe the great virtue of seeing the good in our fellow Jews, in safeguarding their good names. God relies on us to do so whenever we can, especially now in these troubled times. Only G-d knows why He chooses to give us these trials and tribulations; our job is to tell Him, "We don't see any reason to be mad."

It's Gotta Get Better than This

Imagining a Perfect Future

Throughout this book we have discussed the Israeli/Palestinian conflict from historical, political, strategic, and spiritual points of view. Different perspectives, each in its own way, calling for different plans of action designed to work together toward a single destiny. And they all revolve around one fundamental question: is the world moving toward the perfection envisioned by our prophets? And how long will it be until the true purpose of Creation will be fulfilled?

The strength to defeat our enemies, overcome our fears, educate our neighbours, and repair what is broken in this world—all these rely on our deeply-rooted faith that there is a happy ending to this story, and that it is non-negotiable. Every Jewish person should know that there is an end goal, and that this goal is essential to God. There is no other option besides *"Geula"*—the ultimate revelation of the oneness of this fragmented world, the elevation from our natural human state to a divine state of wholeness, of unity. Knowing that the grand finale is on its way gives us the inspiration we need to stand firm and join in the journey, with our own acts of goodness and kindness.

For those who do not see and believe that the world is created with a purpose, what seems like solving problems is more like putting a band-aid on a bleeding wound. *"It's all futile anyway; climate change is going to destroy the planet; there's no end to the inhumanity of man; all we can hope for are temporary solutions so we can extend our survival a little longer, and make our existence a bit more bearable."*

"The Palestinians aren't going anywhere," many Israelis say in despair. *"We will have to learn to live with these problems until, maybe many years from now, they'll be solved."* For the Jewish people, who will never settle for the endless dysfunction of doubt, this is not a life.

Can you look your children in the eye and promise, wholeheartedly, that everything is going to be just fine? Not just fine, but wonderful? Beyond our wildest imagination? Are you comfortable with the idea that all our stories will have happy endings? Or do you believe (even if you're "religious") that sin, death, and hell are inescapable realities that will never end? That's not a Jewish concept. Torah teaches us that all bad things, including death itself, are only temporary. So relax.

Rabbi Yisrael Meir Lau, a former Chief Rabbi of Israel, once met with the Lubavitcher Rebbe. As they were discussing the situation in Israel, Rabbi Lau asked, in a troubled tone, "Rebbe, what is going to be?" The Rebbe replied, "Do not ask what is going to be. That we already know. Instead, ask what we can do to make the world better today."

We know for certain the climax of the story: *"On that day shall Hashem be One, and his Name will be One."* And we will come to realize the story's moral, when all is said and done: *"Revere God and keep His commandments—for this applies to all mankind."*

The only remaining question is how can we "make the world better today."

How To Bring Redemption

Once upon a time, Jewish people who observed the Commandments and performed *mitzvot* generally did so because they were "religious." In simple terms, that meant growing up and living in a religious community. On a more personal level, the motivation toward religiosity was largely about saving oneself from harsh heavenly judgements, and/or buying oneself a portion in the world to come. This is no longer the prevailing situation. It's become more comfortable to do the *mitzvot* not for some ulterior (though spiritual) motive, but purely for the sake of a deepening sense of connection. Sometimes we're feeling a connection with the diverse Jewish community at large; sometimes we can better appreciate the connection with God. When a non-religious soldier who doesn't even necessarily believe in God is asking to put on *tefillin* or wear *tzitzit* into battle, it means something has shifted in the collective consciousness. Nowadays people are not so interested in the mitzvot because of a personal agenda or a threat. The mitzvah is about the intrinsic value of the experience, for its own sake. Philosophically speaking, it draws us closer to a sublime level of reality that rises above the limitations of logic. More simply put, it's about Jewish people just wanting to do Jewish things. It's like a soul sitting up and raising its hand, saying, *"Present!"*

This is the final phase on the way to redemption, before we take off the masks—our own, our loved ones', our world's— and uncover the perfection that's been hiding here all along.

So get on board, why not? Start doing good because its good, not because it benefits you. Start doing Godly things because you are Godly. You don't need to become someone else. There's no reason to change your identity or your clothes or your zip code. Just do a mitzvah. Begin with one mitzvah, like tefillin or Shabbat candles or eating kosher food, and gradually add more until it becomes easy and natural.

Who Is the Moshiach?

A fundamental aspect of what we've been calling the redemption, or the perfection of the world as we know it, is the arrival of Moshiach, the "righteous redeemer." Jews have been waiting for Moshiach and praying for his appearance every day for thousands of years.

Although it's tempting, we're not supposed to try to figure out who Moshiach is. (First of all, if you're not sure whether it's him or not, it probably isn't him.) Better to begin by learning what's been taught by some of the great Torah sages, such as Maimonides, about his character and his expected achievements. The very essence of what Moshiach brings to the world is clarity, knowledge, and truth—and therefore visibility. So when he does show up, we'll know him when we see him. *Everyone* will know.

The pivotal thing we need to know about Moshiach is that if we do our part, he will come. What then is our part? As we have seen throughout this book, true morality means doing what God wants of us. All other forms of what passes for morality are fickle, corruptible, even dangerous. Man-made morality always ends up one way or another in cruelty. Maybe it's time to redefine the idea of morality as "God's will." By knowing and performing God's will, we'll naturally be practicing morality. They go hand in hand. And God's will also brings infinitely more: when the Creator is happy with his universe, everyone in the universe is happy. How do we know what God wants? Ask a Jew who studies Torah—the original, authentic Torah, as given on Mount Sinai, with its written and oral traditions. Profound, ethical, inspiring, and true. There has never been a need for a revised edition to suit our personal predilections. Nor have we ever needed to use it to bully, exploit, or intimidate. Our mission, our privilege, has always been to learn what God needs from us, and to convey that to the rest of the world.

With all that handled, who needs war?

"The Land of Israel has been gifted to us by G-d in our generation with open miracles. There is no doubt whatsoever that we are forbidden to touch it. No one has the right to touch it because it is ours as an everlasting inheritance! The Land of Israel belongs to every single Jew throughout the world, to every Jew of all previous generations, as well as to our children and grandchildren, given to us by G-d..."

—THE REBBE

Learn more about Rabbi Manis Friedman and his work

Rabbi Manis Friedman regularly speaks and teaches at events worldwide. The most popular rabbi on YouTube, he offers free online tips and resources as well as ongoing releases on a wide range of topics, including spirituality, life's challenges, emotional well-being, relationships, dating, marriage and family, and much more.

Sign up for Rabbi Friedman's free email newsletter to get exclusive content, updates, and special offers at:

www.itsgoodtoknow.org

Stay connected:

- itsgoodtoknow.org
- info@itsgoodtoknow.org
- facebook.com/manisfriedman
- youtube.com/manisfriedman
- twitter.com/manisfriedman
- instagram.com/rabbimanisfriedman

ELAD BEN ELUL

Elad Ben Elul is an anthropologist, lecturer, and teacher of Chassidic philosophy. He has a BA in media and journalism from Westminster University and an MSc in digital anthropology from University College London. Elad Ben Elul conducted two years of anthropological fieldwork in West Africa as part of his PhD dissertation in the department of sociology and anthropology of Tel Aviv University. He was granted the first prize for the best PhD dissertation of the Israeli Anthropological Association, later published as the book *Urban Ghana and Privacy in the Digital Age: An Ethnographic Exploration.* Dr. Ben Elul was deeply influenced by Rabbi Manis Friedman's YouTube lectures and, after watching them for a year, decided to write the first Hebrew book based on Rabbi Friedman's teachings, *I Didn't Ask to Be Born.* It became an immediate bestseller, inspiring Jews from all sectors of Israeli society. Dr. Ben Elul and Rabbi Friedman give regular online video lessons in Hebrew and English and present a web series called *The Psychology of the Future.* Their upcoming book, *Back to Innocence,* focuses on gender and intimacy. Dr. Ben Elul is also teaching Tanya and Chassidus to help people connect to their life's purpose.

To learn more about Dr. Ben Elul's work,
visit www.eladbenelul.online.

RABBI MANIS FRIEDMAN

Rabbi Manis Friedman is a world-renowned author, teacher, and speaker, well known for his provocative and incisive wit and wisdom. His international speaking tours, seminars, and retreats take him around the world, and he has been featured on CNN, A&E Reviews, PBS, and BBC Worldwide as well as in such publications as *The New York Times, Rolling Stone, Seventeen, Guideposts, and Publishers Weekly.* He is the author of *Creating a Life That Matters: How to Live and Love with Meaning and Purpose; The Joy of Intimacy: A Soulful Guide to Keeping the Spark Alive; Doesn't Anyone Blush Anymore?* and other titles. Rabbi Friedman is the most popular rabbi on YouTube. He is the dean of Bais Chana Institute of Jewish Studies, which he cofounded in 1971, and the founder of It's Good to Know, a nonprofit life-learning foundation based in New York City.

To learn more about Rabbi Manis Friedman and his work, visit ItsGoodToKnow.org.

In loving
Memory
of
Sami
Hajibay

Dedicated by
Samantha
and
Alexander
Hajibay

ACKNOWLEDGEMENTS

We would like to thank Zalman Friedman for initiating and producing this book in such a tight time frame, and to Zalman Friedman (yes, there are two...) for the wonderful design and layout. Our special gratitude goes to Simcha Gottlieb for his exceptional content editing and sub-editing. Above all, we thank the brave soldiers who are giving their lives to ensure the future of Jewish people everywhere. May G-D bless them.

IT'S GOOD TO KNOW

IT'S GOOD TO KNOW PUBLISHING

itsgoodtoknow.org

Printed in Dunstable, United Kingdom

68471899R00051